Amazon Echo and Alexa User Guide:

The Ultimate Amazon Echo Device and Alexa Voice Service Manual Tutorial

By

Joseph Joyner

Table of Contents

Amazon Echo and Alexa User Guide: The Ultimate
Amazon Echo Device and Alexa Voice Service Manual
Tutorial

By Joseph Joyner

First Published, 2015

Printed in the United States of America

Introduction

The world today is making much technological advancement that you cannot even begin to phantom. One of these advancements that have created much awe to people is the robot. It has never been known just to what extent robots can go, starting from withstanding very many severe conditions which a normal human being cannot like extreme temperatures, to going to other greater extents. It is even amazing on its own that a machine can talk, move and perform many other functions that surpass the capabilities of a normal human being. More inventions are being made even more often than before and you can never tell what will happen next. Sooner or later, machines will take over all our operations and almost everything will become automated.

Now, close your eyes and begin to imagine. Picture a device that could answer all your questions provided you knew how to phrase them correctly: A device that could do simple calculations for you including the number of tablespoons in one cup. A device that could tell you the weather today and calculate for you how many more miles you need to run when working out.

Won't that be amazing? Well, stop imagining and go grab yourself an Amazon Echo!

Amazon Echo is a black cylinder, just about the same size as a tennis ball can. It is a voice box that listens to questions and provides quick answers. However, this does not render it a normal music player that is voice-controlled. Sooner than you know it could be controlling your every activity around the house. It is a combination of a wireless music player, a voice assistant and a home automation hub.

At first it was sold exclusively to Amazon prime members on an invites only basis. The invites kept increasing over time as Amazon put in more effort in producing a more refined product.

During its launching, the Amazon Echo had limited abilities. It was tied to Amazon's music service and allowed users to play songs from the users' own playlists. It also allowed the company's members only to enter voice commands. It could even tell you the weather and answer some basic fact questions. Other than that, it was largely limited. It was also equipped with a few cute extras which were worth the discount price offered.

It seems like every week new features are being added to the Amazon Echo. It has now been released to be openly purchased and as you take home your Echo, you can be sure that there is a bright and broad vision for the Echo.

Chapter 1. Amazon Echo and Alexa

Amazon has made the Echo available at their site and it also provides shipping services.

It has also made the device available to third parties. This provides a platform for the exploitation of each and every inch of the Echo's potential for example the free Alexa Skills Kit. Alexa is the name that the company has given to the voice assistant inside the Amazon Echo. The kit consists of the following:

1. It consists of a device maker which has a sprinkler system connected to the internet. The sprinkler system may also be integrated with Alexa. This enables a user to give voice commands like, "Alexa, please water my garden for 20 minutes".

2. It also contains a surf-report provider. This is used to create new skills for Alexa which provides more options for her. For example, the user can ask Alexa for their latest conditions during their break by telling, "Alexa, ask Surf Status for my forecast".

3. If the third party is involved with making vacuums, they can create a skill for Alexa where the user can enter in a voice command such as, "Alexa, ask the vacuum to start cleaning the guest room" or "Alexa, tell the vacuum to start cleaning the bedroom".

4. The fitness service may also add a skill in the Alexa Skill Kit such that the user feeds in information about his working out history and then Alexa can calculate the number of miles the user ran. This can be done simply by saying, "Alexa, you can ask your fitness to monitor the miles of your running?" or "Alexa, ask My Fitness how many miles I ran last week?

5. The fashion service can also add a skill. For example, they can feed in information about the appropriate outfits for some specific functions. Then the user can get it by asking, "Alexa, ask My Fashion what is the appropriate outfit for a job interview for ladies."

6. The shoe industry can add a skill that enables a user to know their appropriate shoe just by feeding in their shoe size. For example "Alexa, ask My Shoes the appropriate lady shoes for a size 6."

7. Supermarkets can use Alexa to make shopping easier. For example, they can add a skill that enables the user to add items to their shopping list via voice commands. For example, "Alexa, add 2kg of sugar to My Shopping List".

8. Parents can take advantage of Alexa's skills to help their kids with their homework. This is possible because Alexa can answer virtually any question provided it is phrased the in the right way. One amazing feature about Alexa is that if it is not in a position to answer your question it will automatically paste it on Bing (and only Bing) then search it and provide the answer! If you doubt this you can try it for yourself, just ask it, "Alexa, what is probability?" or "Alexa, who is the first president of the United States of America?"

9. The Amazon Echo can also tell you the weather. For example, you can try asking it, "Alexa, what is the weather today?" after sometime you will receive a reply, "Right now, in Los Angeles it is 22 degrees with cloudy skies. The weather forecast of today is mostly sunny, with a high of 27 and a low of 18." It is amazing how accurate and specific Alexa can get sometimes.

This will also help you know how to dress up for the day. Alexa will help you avoid situations where you are wrongly dressed for the weather as it mostly happens to the ladies.

10. It can also be used to tell the time and to set alarms and timers. She has a robotic but very friendly voice. Though she still has a long way to go before becoming fully functional in a lot of situations but there is still hope. It always feels amazing when interacting with the Amazon Echo since it just has very many abilities and it is efficient in its activities and when carrying out commands. It is also encouraging to know that whenever you need something done, all you have to do is ask.

However, the developer tools in Alexa are not just used in the Amazon Echo. It has been proven that they can be used in other third party hardware. The user base that has already been installed targets a larger audience. With the Alexa Skills Kit, programmers can see a very huge potential since they can integrate it with virtually any device that can be connected to the internet ranging from home security lights to heating systems among others.

Chapter 2: What Amazon Echo Can Do

Since its launch, the Echo has been equipped with very many features including both the useful and the less useful ones. For example, it has the feature called 'Simon Says' which enables it to repeat every word you are saying. One thing that is constant is that all features, whether useful or not are easier to use and some of them can actually solve your problems.

We can be allowed to call Alexa a "brain" that is cloud based. Due to this, it is able to update itself since it is connected to the internet. Amazon has continuously added new features to an extent that even the prime purpose as to why the Echo was created (which was to be a music device) has been largely improved. This has been done by linking it to a number of apps such as Prime Music, Pandora, Amazon Music, Audible iHeartRadio, and TuneIn among others.

Most promising features of the Amazon Echo

1. Belkin Wemo and the Phillips Hue products

These two features have been integrated in the Amazon Echo in order to perform many amazing skills around the house on your behalf. For example, they enable the Amazon Echo to turn on the bedside lamp before you get out of bed, or to turn on the fan without you getting up from your favorite chair. They can also be used to dim the lights as you prepare to watch your favorite movies from the couch. All these will be done without you lifting a finger. These two features turn your house into a connected home. This is the feature that will make the Amazon Echo end up in the living room of most people.

2. Google Calendar

The Google Calendar has been integrated in the Amazon Echo to take care of your schedule. You can be able to quickly access your schedule just by asking your Amazon Echo. It is in a position to let you know if you will be available for dinner with friends on a particular day. All you have to do is feed in your voice command and that's it. You can also use it to save any changes in your schedule for future references.

3. It can be used to keep track of the sports scores and their schedule

You and your friends might be super fans of a particular sport (or particular sports). Ever been out maybe drinking and then your friends start to discuss the scores of the game but you can't contribute since you missed it? Yeah, most of us have been there too. However, with your Amazon Echo at hand, you will never have to worry about missing a game. If you want to know the scores of any particular game, all you have to do is ask! This feature is able to provide to you the scores of the game and also give you the full schedule of the sport including when it is scheduled to be playing next. Isn't that just amazing?

4. Traffic

If there is a place where most people lose a lot of time and there is nothing they can do about it, it should be in traffic. Whether you are working or just going out for a ride, traffic affects you in one way or another. The worst case scenario would be when you are going out for a job interview or an important meeting, or it's your first day at work and you are running late because you have been caught up in traffic.

Believe it or not, the Amazon Echo takes care of this too. All you have to do is to save in your work address and Alexa is in a position to let you know the quickest route to work. It can also calculate the travel time so you choose for yourself the option that best suits you. It also provides the option whereby you can choose to include a stop. For example, if you have to pick up a child from school or if you have to drop off some mail or pick up a cup of coffee, just include the address in your voice command and the Echo will do the math for you.

5. TuneIn

This feature enables you to access customized news from many more sources including news Flash Briefing, NPR Business, ESPN Radio and TMZ among others.

6. Making orders

Alexa can also allow you to re-order some products that are Prime-eligible though this may be a little bit challenging to do without having to click the associated app.

Chapter 3. Setting Up Amazon Echo

Setting up the speaker is a little bit complicated since unlike other speakers, you cannot set it up over Bluetooth. You have to set it up over its Wi-Fi network. The Wi-Fi network is the one that give sit access to all the information needed. Without Wi-Fi connection then your Amazon is less useful. It fully depends on the internet when carrying out its tasks. It also takes a few minutes every time you move into a new Wi-Fi network.

Chapter 4. Amazon Echo Advantages

The Amazon Echo is most reliable because it is always connected to a power source so you never have to worry if it has enough charge or not. To some extent, this can be a disadvantage due to portability; meaning that sometimes you cannot use it while on transit unless you have a power source nearby.

As stated earlier, if it does not have the answer to the question that you asked, it will go an extra mile and search for it on your behalf. Therefore to an extent we can conclude that it has the answer to virtually every question as long as you phrase it the right way.

Amazon always makes sure that it is updated constantly, so you can be sure that the information you are getting is up to date and reliable.

Some apps that have been installed are aimed to entertain you. So after having a long day, you can count on your Amazon Echo to help you relax and refresh your mind. When you are busy and your child is crying, or when you just want to keep your child occupied, you can use apps such as "Simon Says" and others to engage or distract your kid until you finish up your work.

It saves you a lot of time because you can ask it perform some tasks for you when you are caught up by many chores. It also saves you the time and money for buying a home automation hub.

Alexa knows where her name came from! Not so many people know where or who they were named after but for a machine to know where its name came from; that is just incredible. Alexa knows that she was named for the library of Alexandria which was used to store the knowledge of the ancient world.

She can also understand complicated questions and help you out with your spelling. Just ask her, "Alexa, how do you spell the word 'conscious'?" she is always quick and ready to help. It is always amazing how she can understand some commands like turning the volume up or down. It always feels like magic just watching her perform the tasks.

It has the added advantage of being able to buy a song for you. For example, when it is playing the 30 seconds preview of "Thinking Out Loud", just say "Alexa, buy that song" and voila! It will be added to your Prime library. You can also control your playlist from the comfort of your couch just by saying, "Alexa, play something else."

The Amazon has an app (the Echo app) which is much like a store for all your history of activities and to do list that you have used on Alexa. It also acts as a user guide on how to talk to Alexa and how to issue your voice commands. The list are created by Alexa and you can access them anytime you are having a problem with issuing a command or when you want to learn more about making Alexa more friendly.

The Echo gives the most successful and most satisfying answer when you speak freely to it and when you trust it than when you have to think about what you are saying.

The good news about Alexa is that it does not have problems with its hardware. It has a speaker that is fully functional and the microphone is good enough to record clear audio that is up to 30 or 40 feet away. It stores every question with a brief audio clip of your voice command on the History screen in the Echo app. You can scroll through and listen to how exactly you asked the question. This will help you correct yourself in case you gave out the wrong voice command.

Amazon has also gone a step further to enable Alexa to ask you if the Amazon Echo heard you correctly. Therefore you are in a position to repeat yourself in case you were wrong, or to repeat your voice command in case you were not audible enough.

The Amazon Echo is a good music player due to its added bonus of a built-in voice control that is excellent. Its functionality has also been greatly improved by the addition of some other apps like weather, traffic, and news features among others.

When the sweet turns sour

Everything has its pretty side and its less pretty side; and Alexa is not an exception. One thing about Alexa that stands out is that you have to be very specific when giving your voice commands. (Just a tip: never ask her to just play songs) I can say that to an extent, the phrase garbage in garbage out (GIGO) applies when speaking to Alexa. If you just ask her to play songs without specifying, she might give you the shock of a lifetime. When you give her general orders she will give back general answers. Like in the case above, if you did not specify the song you want her to play; she will play any song from any category including the songs that annoy parents.

Another thing to take note is that when addressing her you should always start with the name "Alexa" or "Amazon" could also be another alternative though the former (Alexa) is better. If you fail to do this the Amazon Echo will not read your command.

If it does not hear you it just does not. Sometimes it may not be in a position to interpret your command. For example you may have asked it to turn on the sprinklers for 15 minutes, if in case you would like it to stop before the 15 minutes are over, sometimes you might say "Alexa stop!" but it may not.

Sometimes she may fail to know some answers to basic questions. For example, she may give you the answer to the name of the first president of the United States of America but she may not know the name of the second president of the same United States of America. When it comes to her, if a command fails to register, it just fails and there is nothing you can do about it. The only feedback that it gives to show that it is working is a blue green light. Other than, it is difficult to tell if it is fully functional or not.

The fact that the Amazon Echo fully relies on the Wi-Fi connectivity could also be a disadvantage since its abilities are limited if you move to a location with no Wi-Fi network. It is also not advisable to use in places where there is a problem with power connectivity and power surges.

It is also important to note that soon as Alexa starts playing music at any audible volume, it becomes as good as any other normal speaker. No matter how loud you shout a voice command at it, it will not respond and by the time you get up from your couch and turn down the volume, the fun is already gone.

Alexa is good when searching for music but not when playing it. The music coming out of it sounds shallow, compressed and tiny; you can get a much better off speaker at the same price.

A problem may also arise when watching or playing FIFA and the commentators called out the name of Alexis Sanchez who is a striker from Chile. Alexa would be turned on and get ready to receive a command.

Chapter 5. Future of Amazon Echo

The Amazon Echo is already a jaw dropping invention on its own and Amazon is already doing a great job trying to improve its features and exploit its full potential. There isn't that much to recommend but they should really improve on its accuracy. It should also be made more powerful by including more useful apps in it.

Right now, Amazon is slowly introducing it to its customers and it is still being purchased by invitations only. It is most exciting on the first few days when you just want to try it out by asking it a few questions here and there but once all this is done, the 'honeymoon phase' ends and your Amazon Echo isn't exciting anymore. It becomes only useful for performing smaller tasks or just for use when listening to music or podcasts.

I will not fail to recognize that Alexa still holds much potential and every time I am working with her I can't stop wondering what if? What if her skills could be synced up with your daily activities such that when you ask for news, it provides you with news from your favorite magazine? What if there was a way for it to fit into your daily routine instead of you trying to fit yourself into its available features and creating new ones that favor your needs.

Amazon's marketing abilities and some of the apps installed in the Amazon Echo are enough to convince you to purchase one. Most people will purchase it largely because of the home automation abilities even though that is not its primary purpose. Amazon has speculated its worth to be up to $ 2 trillion in the near future and many investors are rushing to get a piece of its shares while stocks last.

Each and every day, new technological advancements are made and it is difficult to keep track of all the changes taking place all around us. The Amazon Echo is one such advancement and the fact that it is still developing should be enough to keep an eye on it. If this is just the beginning yet it can carry out many tasks like holding a steady conversation with someone, what about the near future? It is just amazing how this device is able to interpret commands and other instructions and provide the necessary feedback.

It is true that it has many limitations right now, but this should not be used to discourage its developers as there is still much more room for improvement. Since it has already been made available to third party developers, I would encourage more investors to consider maximizing its full potential since it is an amazing device.

One of the most sure ways of improving on its abilities would be the use of more apps since apps have the potential to perform virtually any function and isn't this what we want for the Amazon Echo? In these days more and more apps have been developed to increase the efficiency of most processes. For example there is an app for boosting the working speed of a machine, an app for waking you up in the morning, an app that helps you monitor your diet and monitor your exercise regime, if all these could be included in the Amazon Echo (the ones which have not been included yet), this could largely increase the functionality of Alexa.

The Amazon Echo has unlimited abilities and unlimited potential. It is very efficient in its abilities and functioning and Amazon is working hard to try and personalize it so as to suit the needs of most of its customers.

I can confidently say that there is a future in the Amazon Echo and sooner than you know it will be in everyone's living room. I even have a feeling that sooner or later the Amazon Echo could become the smartest thing in your living room. It will virtually become your personal assistant in no time. Till then you better hurry and grab your own Amazon Echo while stocks last!

Thank You Page

I want to personally thank you for reading my book. I hope you found information in this book useful and I would be very grateful if you could leave your honest review about this book. I certainly want to thank you in advance for doing this.

If you have the time, you can check my other books too.